THE MOSQUITO

BY LISA OWINGS

BELLWETHER MEDIA • MINNEAPOLIS, MN

Jump into the cockpit and take flight with Pilot books. Your journey will take you on high-energy adventures as you learn about all that is wild, weird, fascinating, and fun!

This edition first published in 2013 by Bellwether Media, Inc.

Library of Congress Cataloging-in-Publication Data

Owings, Lisa.
The mosquito / by Lisa Owings.
pages cm. – (Pilot. Nature's deadliest)
Audience: 8-12
Includes bibliographical references and index.
Summary: "Fascinating images accompany information about the mosquito. The combination of high-interest subject matter and narrative text is intended for students in grades 3 through 7"– Provided by publisher.
ISBN 978-1-60014-881-1 (hardcover : alk. paper)
1. Mosquitoes as carriers of disease–Juvenile literature. 2. Malaria–Juvenile literature. I. Title.
RA640.O95 2013
614.4'323–dc23

2012039832

Printed in the United States of America, North Mankato, MN.

CONTENTS

A Mysterious Illness

Mississippi teen William Crasto was just 11 years old when his life changed forever. He and his family had gone on a fishing trip to Louisiana in the summer of 2007. The trip had been great except for the **swarms** of mosquitoes. It was shortly after the Crastos returned home that things started to go terribly wrong for William.

A fever and pounding headache kept William home from school for a few days. Soon a mysterious rash spread over his body. Then he began vomiting. William's illness made him so tired he could barely stay awake. His parents were worried. They hoped William's flu-like symptoms would improve. Then William lost his sense of balance. That's when the Crastos knew something was seriously wrong.

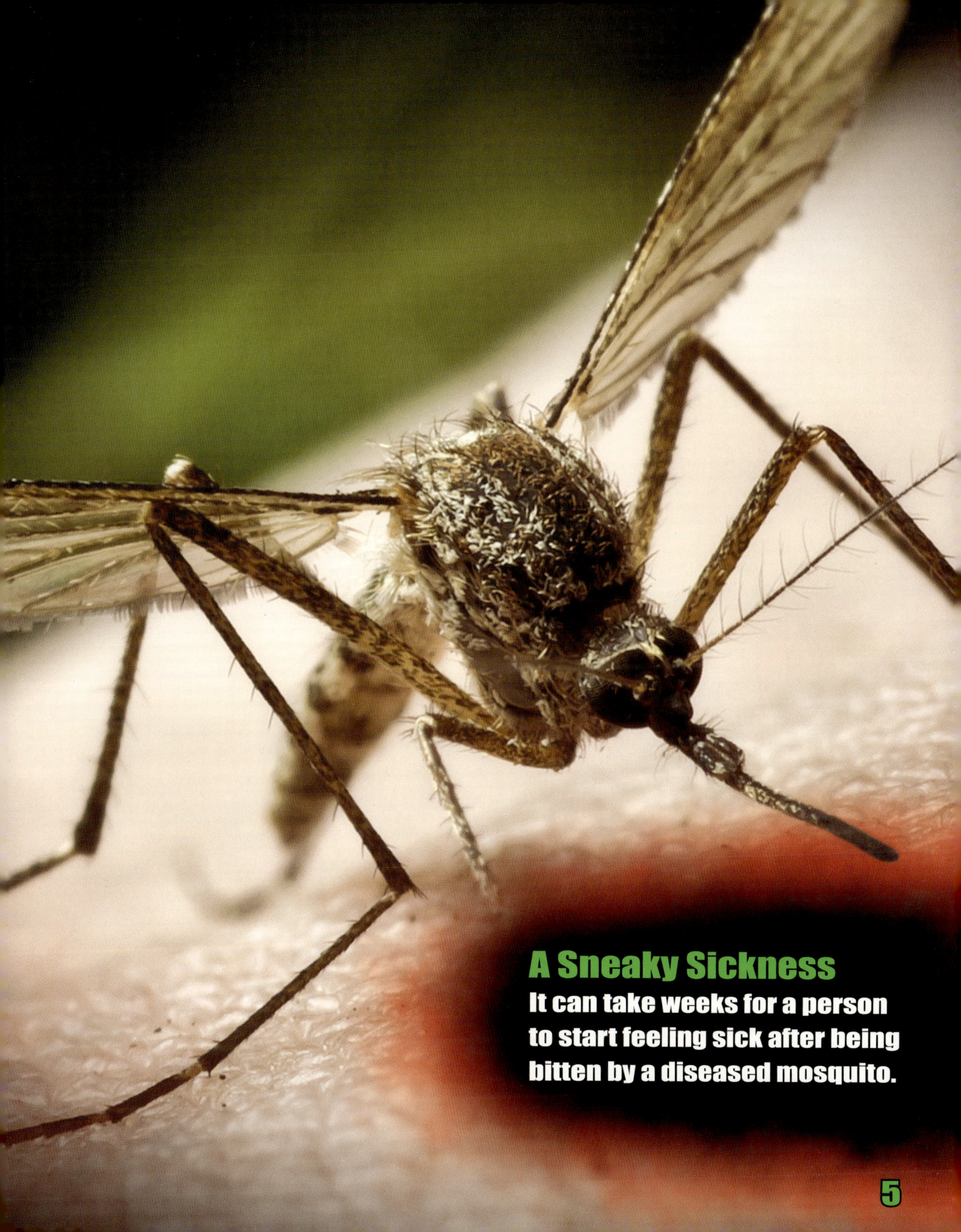

A Sneaky Sickness

It can take weeks for a person to start feeling sick after being bitten by a diseased mosquito.

William's parents rushed him to the hospital. By the next morning, he had lost feeling in his back and one of his legs. He couldn't even get out of bed. One day, William almost did not wake up. His doctors ran test after test. It seemed they would never find the cause of William's suffering.

William was eventually sent home in a wheelchair. He returned to the hospital each week for blood tests. It would be three long months before the Crastos got the news they had been waiting for. The doctors finally discovered what had made William so sick. It was nothing more than a tiny mosquito.

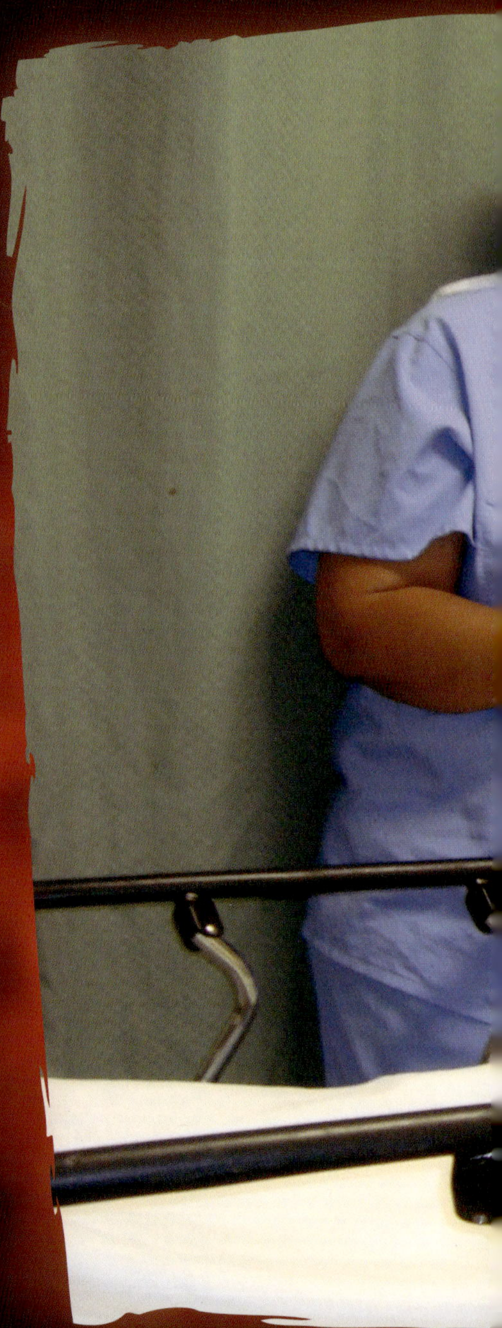

Are You at Risk?

Most people's bodies can fight off diseases carried by mosquitoes. They might not even know they are sick. Young children and elderly adults are most at risk of having severe symptoms.

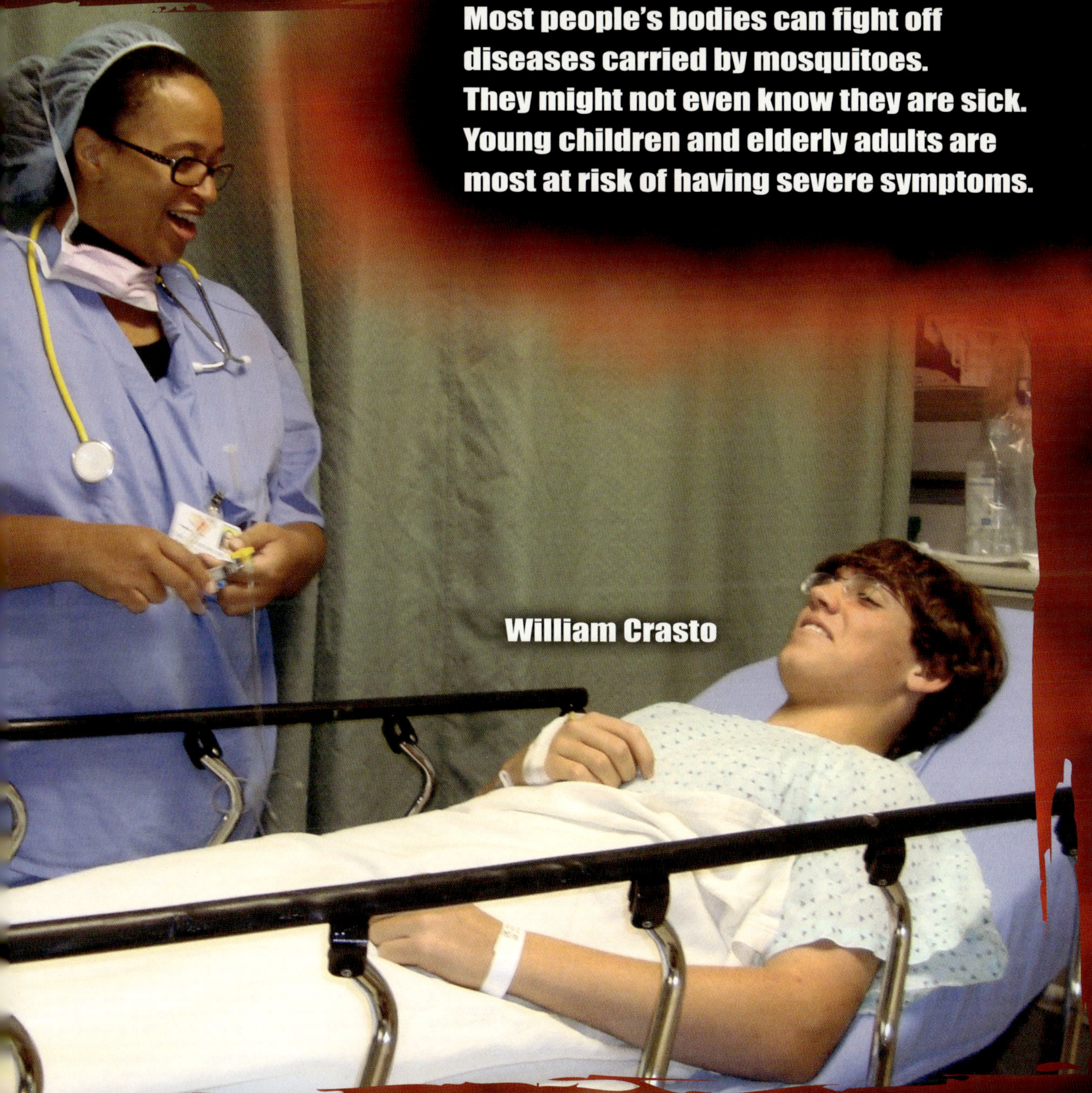

William Crasto

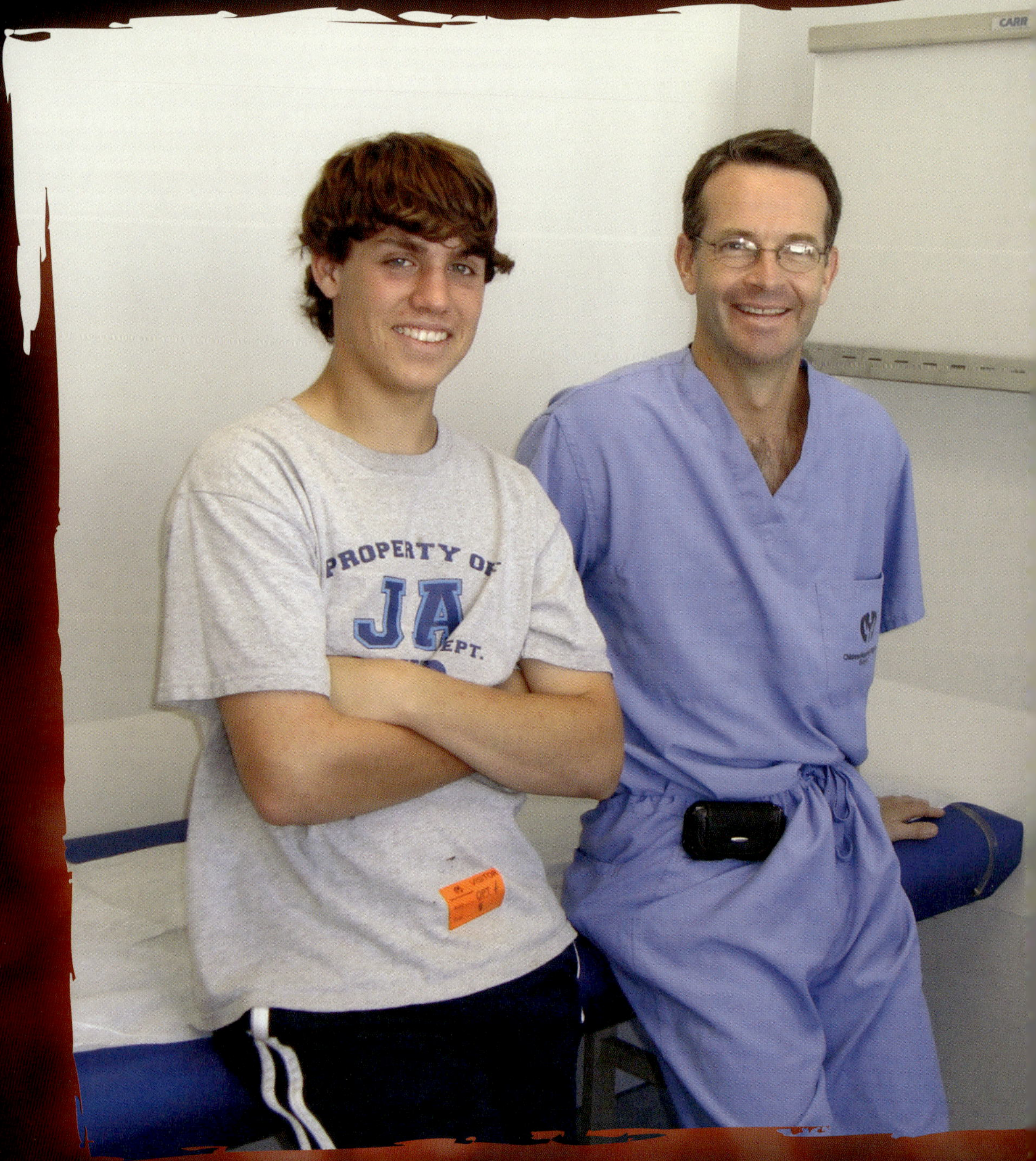
PROPERTY OF
JA

West Nile in New York

The West Nile virus was first discovered in 1937 in Africa. It reached the United States in 1999. New York City experienced a frightening outbreak that year. It put around 60 people in the hospital. Seven of the patients died.

William had been **infected** with the West Nile **virus**. The disease had caused his brain and the tissues around his **spinal cord** to swell. Such a severe response to the virus is often a death sentence. William was lucky to be alive, even if he could no longer walk.

William faced years of **physical therapy**. By age 16, he was doing more than just walking. He was cross-country running. He was also spreading the message that mosquitoes can do more than just make you itch.

Deadly Diseases

Mosquitoes kill more people than any other animal on Earth. They don't have razor-sharp teeth or **lethal** poison. Instead, they spread disease. Most of the diseases they carry are caused by viruses. The deadliest **mosquito-borne** disease is malaria. It is caused by a **parasite**. Each year, mosquitoes infect more than 700 million people with life-threatening diseases. More than 2 million of those infected die.

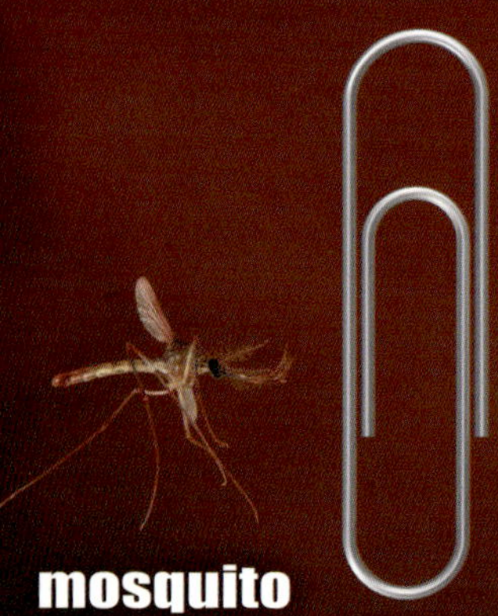

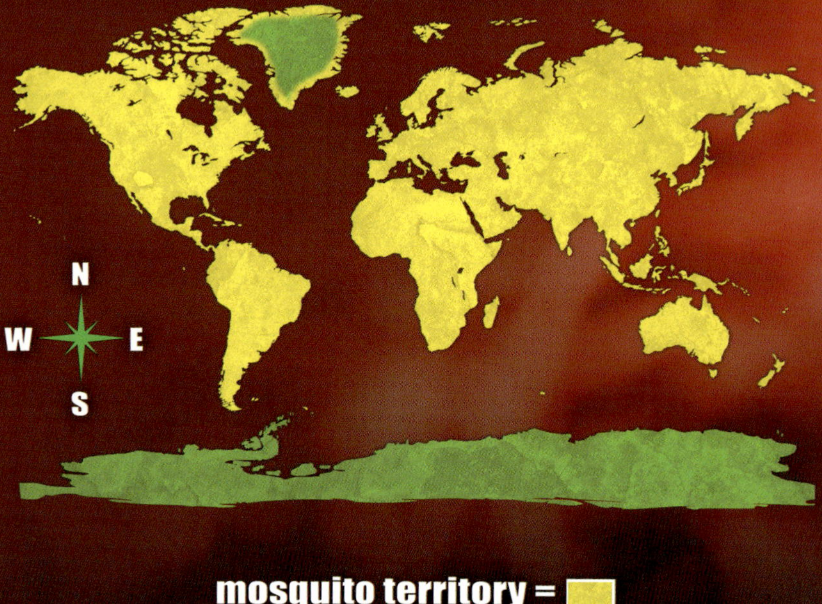

Mosquito-Borne Diseases and Symptoms

Disease	Primary Region	Symptoms	Complications
Malaria	Africa, Southern Asia, South America	fever, chills, sweating, vomiting, headache	brain swelling or damage, difficulty breathing, organ failure
Yellow fever	Africa, South America	fever, muscle aches, vomiting, dizziness, yellowish skin and eyes, bleeding from eyes, mouth, and nose, irregular heartbeat	liver failure, kidney failure, seizures, coma
Dengue fever	Southeast Asia, South and Central America	fever, headache, pain in muscles, bones, and joints, rash, vomiting, bleeding gums or nose	lung damage, liver damage, heart damage
Encephalitis (swelling of the brain)	worldwide	severe headache, fever, sleepiness, weakness, confusion, seizures, paralysis or loss of feeling, trouble seeing, hearing, or speaking	difficulty breathing, coma, brain damage
West Nile virus	worldwide	fever, headache, body aches, sleepiness, rash, stiff neck, loss of balance, sudden weakness, paralysis	coma, long-term muscle weakness, long-term paralysis, brain damage

Mosquitoes are known for biting other animals and sucking their blood. However, it is only the female mosquito that does this. She needs a blood meal to lay her eggs. Sometimes the animal she bites is infected with a disease. If the disease can survive in the mosquito's body, she may infect the next animal she bites.

A female mosquito is attracted by certain chemicals in an animal's breath or sweat. She can detect these chemicals from up to 100 feet (30 meters) away. She can also sense an animal's body heat. Once she has chosen her victim, she lands on the animal's warm skin. Then she pierces it with her long mouthparts. Soon she reaches blood. She drinks greedily, sucking the liquid into her belly.

Something Sweet

Male mosquitoes don't drink blood. They feed on nectar. This is a sugary liquid produced by plants.

The mosquito's **saliva** flows into a bite as she drinks. It keeps her victim's blood from getting too thick. This saliva is what makes the mosquito an enemy of humans. It causes an **allergic reaction** in people. The victim's wound begins to itch within seconds of being bitten. This annoying sensation usually goes away within a few days. But for some, it is just the first sign that a terrible disease is attacking their body. A person who is bitten by a diseased mosquito can become deathly ill. Symptoms can be treated, but there is no cure for most mosquito-borne illnesses.

At-Risk Animals

Mosquito-borne diseases don't just affect humans. The West Nile virus also infects birds. Monkeys can catch yellow fever, and horses often suffer from encephalitis.

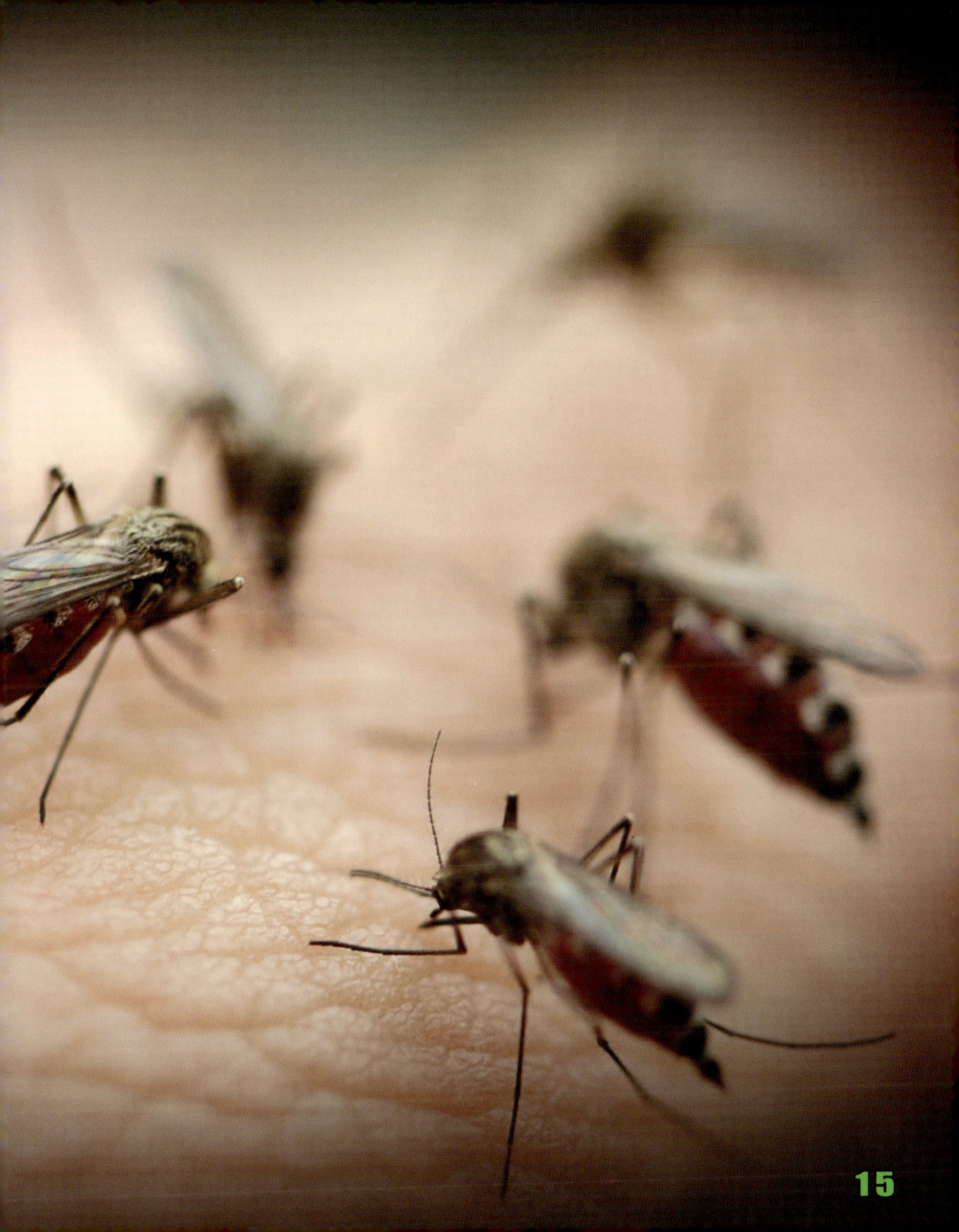

Mosquito Attacks

Not all female mosquitoes carry diseases. Only the ones that have had a blood meal from a sick animal do. A female mosquito needs a blood meal when she wants to lay eggs. If the blood she drinks contains a disease, she can spread that disease during her next meal.

No Water, No Eggs

You can keep mosquito numbers down by draining standing water near your home. This takes away places where mosquitoes can lay eggs.

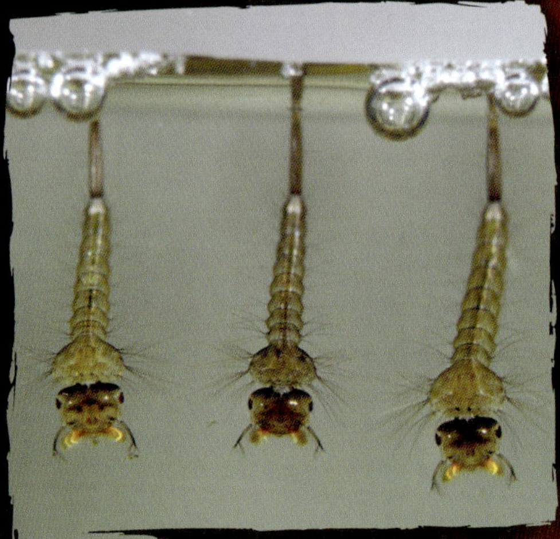

Many scientists believe mosquito-borne illnesses will become more common in the future. **Climate change** allows tropical diseases to survive in new regions. It also creates more warm, wet areas where mosquitoes can reproduce. The rising number of mosquitoes makes it difficult to prevent the spread of mosquito-borne diseases.

The best way to avoid being infected by a mosquito is to prevent mosquito bites. Try to stay indoors around sunrise and sunset. This is when mosquitoes are most active. Be sure to fix any holes in your window screens so they can't get inside. When outdoors, use bug spray. Bug spray with **DEET** is the most effective. However, some people are sensitive to the chemical. Wearing long, light-colored clothing also helps protect against bites. Dark clothing attracts mosquitoes.

If you are bitten by a mosquito, wash the bite with soap and water. Use ice or anti-itch cream to stop the urge to scratch. Seek medical attention if you experience a fever or other severe symptoms over the next few days or weeks.

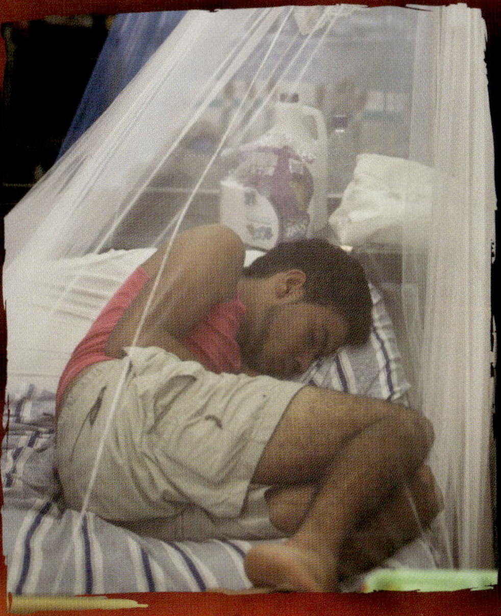

Night Bites

People in areas where malaria is common fight the disease by using bed nets. These nets have openings that are too small for mosquitoes to get through. They are also often treated with chemicals that kill mosquitoes.

Tropical Travels

Tell your doctor if you are planning a trip to a tropical area. There are medicines that can protect you from certain mosquito-borne illnesses.

Attack Facts

- Malaria kills more than 1 million people every year.
- Malaria is the second-deadliest disease in Africa after HIV/AIDS.
- More than 4,000 people in the United States were diagnosed with West Nile virus in 2012. There were only 712 cases in 2011.

Controlling mosquitoes is an important step in keeping people healthy. However, chemicals used to control mosquitoes can be harmful to people, animals, and the environment. Scientists are working to find safer ways to reduce the mosquito population. Doctors are also developing medicines to protect against or treat certain mosquito-borne illnesses.

Someday those itchy red bumps may be the most we have to fear from mosquitoes. Until then, we have to learn to live with mosquitoes while protecting ourselves. Next time you hear that buzz in your ear, steer clear!

Glossary

allergic reaction—a negative physical reaction to a substance

climate change—the gradual warming of Earth and other changes in weather

DEET—a chemical that makes it harder for mosquitoes to sense humans

infected—had a disease introduced into one's body

lethal—deadly

mosquito-borne—carried and passed on by mosquitoes

parasite—an animal or plant that lives on or inside another animal or plant

physical therapy—the treatment of injured or damaged body parts through exercise, massage, and other techniques

saliva—the fluid in the mouth of a person or animal; a mosquito's saliva can pass diseases to people and animals.

spinal cord—a cord of tissue that runs from the brain down the spine; the spinal cord allows the brain to communicate with the body.

swarms—large groups of insects that move together

virus—a tiny organism that lives and multiplies inside cells; viruses cause many diseases, including the common cold.

To Learn More

At the Library

Jurmain, Suzanne. *The Secret of the Yellow Death: A True Story of Medical Sleuthing*. Boston, Mass.: Houghton Mifflin, 2009.

Markle, Sandra. *Mosquitoes: Tiny Insect Troublemakers*. Minneapolis, Minn.: Lerner Publications, 2008.

Ollhoff, Jim. *Malaria*. Edina, Minn.: ABDO Pub. Co., 2010.

On the Web

Learning more about mosquitoes is as easy as 1, 2, 3.

1. Go to www.factsurfer.com.

2. Enter "mosquitoes" into the search box.

3. Click the "Surf" button and you will see a list of related Web sites.

With factsurfer.com, finding more information is just a click away.

WWW.FACTSURFER.COM

Index

The images in this book are reproduced through the courtesy of: Kallista Images/Getty Images, front cover, pp. 12-13; Dmitrijs Bindemanis, pp. 4-5; Potapov Alexander, p. 6; Crasto family, pp. 6-7, 8-9; sydeen, pp. 9, 10; Natalya Filimonova, pp. 14-15; Age Fotostock/SuperStock, p. 16; Darren McCollester/Stringer/Getty Images, p. 17; Corey Rich/Getty Images, p. 18; Reuters/Edgard Garrido, p. 19; Thomas Northcut/Getty Images, p. 20.